30th Annual Saudi-Japan Symposium on Technology in Petroleum Refining and Petrochemicals 2021

Online
13 – 14 December 2021

ISBN: 978-1-7138-4206-4

Printed from e-media with permission by:

Curran Associates, Inc.
57 Morehouse Lane
Red Hook, NY 12571

Some format issues inherent in the e-media version may also appear in this print version.

Printed with permission by Curran Associates, Inc. (2022)

For permission requests, please contact King Fahd University of Petroleum & Minerals
at the address below.

King Fahd University of Petroleum & Minerals
Dhahran 31261
Kingdom of Saudi Arabia

Phone: +966 (13) 860 0000

info@kfupm.edu.sa

Additional copies of this publication are available from:

Curran Associates, Inc.
57 Morehouse Lane
Red Hook, NY 12571 USA
Phone: 845-758-0400
Fax: 845-758-2633
Email: curran@proceedings.com
Web: www.proceedings.com

الندوة السنوية السعودية اليابانية الثلاثون

30th Annual Saudi-Japan Symposium

التقنية في تكرير البترول والبتروكيماويات

Technology in Petroleum Refining & Petrochemicals

Abstracts

10-9 جمادى الأولى 1443هـ

December 13-14, 2021

KFUPM Research Institute
King Fahd University of Petroleum & Minerals
Dhahran, Saudi Arabia

Preface

This is the Abstracts of the 30th annual virtual symposium "*Technology in Petroleum Refining & Petrochemicals*" held on December 13-14, 2021. This event, jointly organized by KFUPM, Saudi Aramco, the Japan Petroleum Institute (JPI), and Japan Cooperation Center, Petroleum (JCCP), brings together researchers from industry and academia to discuss recent downstream technologies related to refining and petrochemicals. The main theme is "Innovation towards Circular Carbon Economy".

At this year's symposium, there are 20 presentations featuring 2 keynote lectures, 18 oral and 10 poster presentations. The topics include recent trends in refining, low carbon technologies, CO_2 utilization, plastic waste recycling, catalytic technologies and processes. Distinguished panel of speakers come from Aramco R&DC, KFUPM research centers/academic departments, KAUST, Idemitsu, ENEOS, JGC, Honeywell UOP and Japanese Universities: Waseda, Tohoku, Osaka, Ibaraki, Kanagawa, Hokkaido and Tsukuba.

We extend our thanks to KFUPM, Saudi Aramco, JPI, JCCP, distinguished speakers and attendees for their active participation and look forward to your participation in our next symposium.

Dr. Hassan S. Alasiri
Director, Center for Refining & Advanced Chemicals
http://ri.kfupm.edu.sa/crp/
KFUPM-Dhahran, Saudi Arabia

Technology in Petroleum Refining & Petrochemicals

Innovation towards Circular Carbon Economy

***Online Symposium (Zoom Webinar):** December 13-14, 2021*

Day One: Monday, December 13, 2021

Each presentation includes 5-minutes Q&A *Program as of Nov. 22, 2021*

Technology in Petroleum Refining & Petrochemicals

Innovation towards Circular Carbon Economy

Online Symposium (Zoom Webinar): *December 13-14, 2021*

Day Two: Tuesday, December 14, 2021

SESSION FOUR CATALYST TECHNOLOGIES ***Chairman: Prof. Teruoki Tago, Tokyo Institute of Technology***

POSTER SESSION ***Moderator: KFUPM***

10:30	**10 poster presentations** (seven min each - *Program on Page 3)*	
11:40	***Prayer & Lunch Break***	

SESSION FIVE CATALYTIC PROCESSES ***Chairman: Dr. Ali Alzaid Aramco R&DC***

Each presentation includes 5-minutes Q&A *Program as of Nov. 22, 2021*

EFFICIENT CO_2 UTILIZATION THROUGH HETEROGENEOUS CATALYTIC PROCESSES

Jorge Gascon

King Abdullah University of Science and Technology, KAUST Catalysis Center (KCC), Advanced Catalytic Materials, Thuwal 23955, Saudi Arabia.

Abstract

The increasing global CO_2 levels has led to a massive thrust in research on both Carbon Capture and Storage (CCS) and Carbon Capture and Utilization (CCU). It has been posited that, in terms of volume, the contribution of CCU will be significantly less as compared to CCS for avoiding CO_2 emissions and achieving the "2 degree scenario" (2DS) goals. However, what cannot be denied is that the immediate economic potential of CCU far outweighs that of CCS especially considering the fact that large scale capital investment is required in case of the latter.

While it is clear that CCS does need to be implemented in order to realistically achieve the 2DS goals, what is not commonly considered is that instead of approaching CCS and CCU as two separate methods of carbon mitigation, we should think of how CCU can, in fact, help address the problems faced for the implementation of CCS. With research intensifying on CCU, it is not impractical to think of an economic cycle where the profits gained from CCU can help to offset the costs of CCS if an integrated system of CCSU (Carbon Capture Storage and Utilization) is implemented. The reuse of stored CO_2 can become beneficial if consumption of fossil fuels is greatly reduced over the next century and the stored CO_2 becomes the chief feedstock for carbon-based chemicals.

In this presentation, we will present several routes based on careful choice of catalytic components and reactor configuration to increase selectivity and productivity in the direct hydrogenation of CO_2 to methanol, light olefins, aromatics and liquid fuels.

Future Trends in Petroleum Industry

Takashi Matsushita

President of JPI, Executive Vice President of Idemitsu Kosan Co.,Ltd., Japan

Abstract

Future scenario of Primary energy by fuel in the worldwide and Japan published from EIA and Japanese government will be expounded. Idemitsu also set up Long term energy business scenarios. Most of these scenarios suppose that the crude oil contribution to energy supply toward 2050 would be reduced gradually in order to respond to Global warming. Whereas, energy security i.e. stable energy supply to the society is also very important mission of Petroleum Industry, so smooth energy transition from Fossil fuel to renewable energy, biomass, synthetic fuel, hydrogen, ammonia etc. must be necessary. One of important key points for our success to realize it with economic rationality is "Technological Innovation". The areas we all need to tackle together beyond individual state, industry, government, academia frameworks will be discussed.

Technology Challenges and Opportunities in Carbon Capture and Utilization in Process Industry

Aqil Jamal, Ph.D.
Chief Technologist Carbon Management Division; Research and Development Center, Saudi Aramco, Dhahran, Saudi Arabia 31311.

Abstract

Long-term global climate goals may not be attainable without Carbon Capture Utilization and Storage (CCUS). It is widely believed that CCUS is one of the few technologies available that can decarbonize both power generation and heavy industries in an economically affordable way. As countries across the globe set net-zero target, there is a growing recognition that CCUS is essential to fully decarbonize hard-to-abate sectors. Therefore, implementation of CCUS is urgently needed because any further delay will make it harder to achieve the global climate goals. The technology is one of the key enablers for CCUS. This presentation highlights stationary and Direct Air Carbon Capture (DACC) in conjunction with CO_2 utilization technologies that can potentially contribute to balancing emissions between sources and sinks in the circular carbon economy framework.

PREPARATION OF SELF-SULFUR-DOPED ACTIVATED CARBON FROM PETROLEUM COKE FOR ELECTROCHEMICAL ENERGY STORAGE

Syed Shaheen Shah,[1,2] Md. Abdul Aziz,[1],* Zain H. Yamani[1,2]
[1] Interdisciplinary Research Center for Hydrogen and Energy Storage, KFUPM, Dhahran 31261, Saudi Arabia
[2] Physics Department, KFUPM, Dhahran 31261, Saudi Arabia
*Corresponding Author E-mail: maziz@kfupm.edu.sa

Extended Abstract

Due to recent technological developments, energy demands are rising with time. Renewable energy resources provide a considerable alternative to meet global energy demands. But due to its time-dependent operations, a robust storage system is required, which can store vast amounts of energy in a short time. Energy storage has earned a key position in energy devices due to its viability compared to energy harvesting devices. Also, in the last few decades, energy crises forced scientists and researchers to search for novel materials that may have the potential for energy harvesting and storage devices. In this regard, supercapacitors (SCs) based on carbon materials have recently received tremendous attention. A SC is an energy storage device with high energy and high-power densities and can be completely charged in seconds. By carefully selecting electrode materials and using cost-effective and simple preparation methods, SCs can be developed to store renewable energy on a larger scale. Specifically, sulfur-doped activated carbon (SAC) has remained a suitable electrode material for developing efficient SCs with high specific energy and specific power combined with low self-discharging. However, low yield AC due to low carbon content and external sulfur doping cause their high price. Therefore, a byproduct of oil refining, petroleum coke with a high carbon and sulfur content (~90 wt.%), could be considered an excellent raw material for the facile synthesis of sulfur-doped AC with a high yield. Here, we present a high-yield synthetic strategy to prepare highly porous petroleum coke-derived SAC (PCSAC). Following a comprehensive characterization of the prepared PCSAC, using X-ray diffraction, Raman spectroscopy, field emission scanning electron microscopy, X-ray photoelectron spectroscopy, thermogravimetry & differential scanning calorimetry, and Brunauer-Emmett-Teller analysis, its electrochemical performance as electrodes of the SC is investigated. The SAC-based symmetric SC's electrochemical performance exhibited a high specific capacitance with low equivalent series resistance and good energy and power densities. The fabricated PCSAC-based SC also shows high stability and Coulombic efficiency.

Keywords: Petroleum Coke; Activated Carbon; High Yield; Supercapacitor; Electrochemical Energy Storage

Introduction

Nowadays, there is a need for clean energy sources to preserve natural resources and regulating global energy consumption. Supercapacitors (SCs) have gained scientists' attention due to their enhanced electrochemical performance, comprising superior cycling life, high specific power, and a fast charge-discharge rate. Likewise, SCs also exhibit excellent reversibility and power density along with their applications in electrical and renewable devices.[1] Compared to batteries, SCs can manage the power density and the particular energy in several orders of magnitude, making it a viable alternative energy storage system.[2] The SCs consist of electrolytes, current collectors, and electrodes. Up to date, several natural biomass materials have been used as precursors for the production of porous activated carbon (AC) for SCs.[3-6] Specifically, sulfur-doped AC (SAC) have remained suitable electrode materials for developing efficient SCs with high specific energy and specific power combined with low self-discharging.[7] The oil refinery process produces petroleum coke (PC) with a high carbon concentration of ~90% containing sulfur. To prepare AC from PC, physical and chemical activation methods were utilized, where the activation time, agent, and temperature play a vital role in the preparation of porous AC, the surface area, and porosity of AC are affected by these parameters.[7] We used simple pyrolysis using carbonate salt as activating agent to prepare the highly porous PC-derived SAC (PCSAC) in this study. In terms of SCs applications, the PCSAC-based electrodes with low IR-drop performed excellently.

Experimental Procedure

The fine powder of PC (particle size ≤ 100 μm) was mixed with $NaHCO_3$ in a 1:4 ratio (by mass) and heat-treated for 5 hours at 800 °C in a tube furnace with a heating and cooling rate of 5 °C/min under a N_2 environment. Afterward, the carbonized product was washed with 1 M HCl and DI water, filtered, and dried for 12 hours in an electric oven at 70 °C, to get PCSAC. The prepared PCSAC was characterized by various characterization tools, including FESEM, EDS, XRD, Raman spectroscopy, XPS, and BET. To fabricate SC electrodes, 50 mg of the produced PCSAC was mixed with 10 mL DI water and ethanol (50/50 wt.%) and 40 μL Nafion (binding agent) and sonicated for 30 minutes. 200 μL of the solution was dropped onto a steel foil (1 cm^2 working area) and dried in an electric oven at 70 °C for 5 hours. Electrochemical measurements of the PCSAC-based symmetric SC were conducted using a two electrodes cell where filter paper soaked in 1 M KOH electrolyte was used as a separator.

Results and discussions

XRD was used to determine the graphitization degree and crystallinity of the PCSAC. The XRD spectrum of PCSAC (Fig. 1a) revealed two Bragg diffraction peaks at 2θ=25° and 44° representing the C(002) and C(101) planes. The broad C(002) diffraction peak is caused by the structural disorder of graphitic arrangement in the PCSAC that gained throughout the activation and porosity generation processes. The peak 2θ=44° suggested a hexagonal graphite lattice. The broadening of the two diffraction peaks indicated a low degree of graphitization or a high degree of edge expansion.[5] The PCSAC Raman spectrum, shown in Fig. 1b, contains two broad peaks centered at 1320 and 1600 cm^{-1}, representing defective (D) band (resulting from the vibrations of dangling bonds present in the carbon atoms) and the graphitic (G) band (resulted from the vibration of sp^2 bonded carbon atoms present in a 2D hexagonal lattice) of the PCSAC, respectively.[3] The ratio of D to G intensities was determined to be 1.6, which confirms the defective dominant structure of the PCSAC.

Fig. 1. (a) XRD and (b) Raman spectroscopy spectrum of the PCSAC.

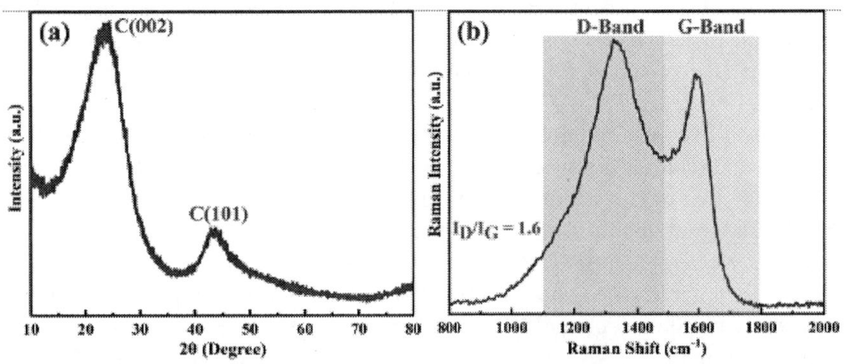

The surface morphology of the PCSAC was investigated using FESEM micrographs. FESEM micrographs of the PCSAC at various magnifications are shown in Fig. 2a-d. The porous features of the PCSAC constructed of microstructures with a network of open pores are demonstrated in these micrographs by combining rough, fractured, and smooth surfaces.[8] The typical EDS spectrum of the PCSAC is shown in Fig. 2e, indicating carbon, oxygen, and sulfur as the major elements. The corresponding EDS elemental mapping of carbon, oxygen, and sulfur represents the even distribution of sulfur in the prepared PCSAC. XPS was also used to investigate functional groups of prepared PCSAC, which confirmed the existence of C1s as a major element with O1s and S2p as minor elements in the prepared PCSAC (Fig. 2f). The C1s deconvoluted spectrum of PCSAC (Fig. 2g) shows the peaks for C=C/C-C (~284.6 eV), C–OH (~286.4 eV), –COOH (~290.1 eV) functional groups.[9] The XPS analyses reveal that C is the major element in the PCSAC, whereas the presence of O is also confirmed (Fig. 2h), which is common in biomass-derived AC.[9] The existence of the S2p at around 169 eV (Fig. 2i) is due to the major content of sulfur in the PC. Figure 2j depicts the nitrogen adsorption-desorption isotherms for the prepared PCSAC. The BET study (Fig. 2j) showed a type IV absorption-desorption isotherm for the porous PCSAC, as well as a large nitrogen adsorption-desorption capacity. The PCSAC shows a larger hysteresis loop from P/P_o=0.38 to P/P_o=0.90, showing the presence of mesopores. Furthermore, a significant increase in the loop in the high relative pressure range (P/P_o=0.78 to 0.9) showed macrospores in the PCSAC. On the other hand, the isotherm displays a significant nitrogen uptake at low pressure ($P/P_o < 0.38$), indicating the presence of micropores. The PCSAC had a BET surface area of ~450 m^2/g and a pore volume of ~0.08 cm^3/g. In Fig. 2k and its inset, the Barrett-Joyner-Halenda (BJH) pore-size distributions for the PCSAC are displayed. The BJH result revealed a wide range of pore widths (ranging from 1 to 200 nm), with an average pore width of ~1.2 nm for BJH adsorption. The presence of dominated micropores in the PCSAC was suggested by these pore sizes and average pore width ranges. The combined behavior in macro-, meso-, and micro- pores helped to improve the specific surface area, allowing for fast ion diffusion at the material surface.

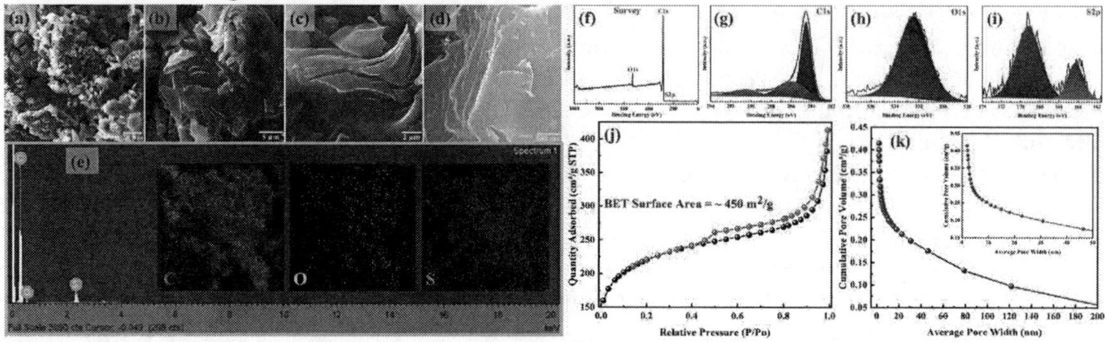

Fig. 2. (a-d) FESEM micrographs recorded at different magnifications and (e) EDS spectrum & elemental mapping of the PCSAC. XPS spectra for the PCSAC (f) survey scan and high resolution deconvoluted (g) C1, (h) O1s, and (i) S2p. (j) The nitrogen adsorption-desorption isotherm and (k) the BJH pore-size distributions (the inset represents magnified region).

The electrochemical performance of the PCSAC-based SC was investigated in a symmetric two-electrode cell in 1 M KOH electrolyte. Figure 3a shows a cyclic voltammogram (CV) of the as-prepared PCSAC at a 10 mV/sec scan rate. The symmetric rectangular shape of the CV with an expanded area demonstrates that the PCSAC-based SC exhibit EDLC behavior. As shown in Fig. 3b, CVs of the PCSAC-based SC were obtained at various scan rates ranging from 10 to 100 mV/s. There is no visible peak from a Faradaic current over the OPW in the PCSAC-based SC, which has a rectangular shape. This suggests that charge storage is caused by the adsorption and desorption of electrolyte ions and that charge/discharge at the EEI is reversible. The CV curves flatten as the scan rate increases, possibly because the electrolyte ions cannot easily access the electrode for charging and discharging at a high scan rate. The charge stored on the electrode, on the other hand, did not increase at a constant rate. This reduced electrolyte ion accessibility to the active material and tilted the CV shape. Figure 3c shows the GCD profiles of the PCSAC-based SC, measured at different current densities from 0.5 to 5 A/g. The PCSAC-based SC displays typical charge-discharge performance with symmetrical triangular shapes and a relatively small IR drop. The specific capacitances were calculated from the CV (at different scan rates) and GCD profiles (at different current densities), and the resultant specific capacitances are shown in Fig. 3d. The PCSAC-based SC exhibited a maximum specific capacitance of 139 F/g at a current density of 0.5 A/g.

As shown in Fig. 3e, CVs of the PCSAC-based SC in the OPWs of 0-0.4, 0-0.6, 0-0.8, and 0-1.0 V maintained their quasi-rectangular shapes suggests that the PCSAC-based EDLC is adjustable to the wide OPWs. The cyclic stability of electrode materials is a critical issue to consider while justifying energy storage performance. As a result, 1000 consecutive GCD cycles at a fixed current density of 10 A/g were used to test the cycling performance of the PCSAC-based SC. For 1000 GCD cycles, as shown in Fig. 3f, the PCSAC-based SC retained ~91% of its initial capacitance while maintaining ~87% coulombic efficiency. The structural breakdown and the electrolyte ions trapped in the pores of PCSAC from repeated charge/discharging are the leading causes of the slight decrease in capacitance.[10] EIS is another critical tool for assessing an EDLC's electrons and ions' transport. The EIS investigation was carried out at 0 V with frequencies ranging from 1 Hz to 1 MHz, and the corresponding Nyquist plots before and after the PCSAC-based SC stability test are shown in Fig. 3g. Semicircles indicate charge transfer resistance (R_{ct}) in the high-frequency portions of the graphs. The almost vertical lines parallel to the y-axis in the low-frequency zones demonstrate optimal capacitive behavior. The solution resistance (Rs), which is a sum of the total electrolytic resistance, electrode material intrinsic resistance, and the resistance between the electrode material and the current collector, is shown by the intersection of the semicircle with the x-axis. Before stability, the Nyquist plot had a R_s value of 30 Ω, showing vital ion transport and conductivity between the electrolyte and the active material's surface. The R_{ct} was measured at 136 Ω, indicating that the material has good charge transfer and conductivity. The R_s value after stability is 30 Ω, the same as before stability, and the R_{ct} value after stability increased to 182 Ω. This demonstrates excellent ion diffusion and conductivity and little material deterioration, even after 1000 continuous GCD cycles, confirming the PCSAC-based SC's stability. The energy and power densities of the PCSAC-based SC were determined from the GCD-specific capacitance values and represented in the form of the Ragone plot, as shown in Fig. 3h. At a power density of 250 W/kg, the PCSAC-based SC delivered an energy density of 19.3 Wh/kg. At a high-power density of 2500 W/kg, an energy density of 9 Wh/kg was achieved. The PCSAC is a suitable electrode material for SCs due to its high specific capacitance, outstanding rate performance, good cycle stability, and its low-cost and straightforward synthesis procedure.

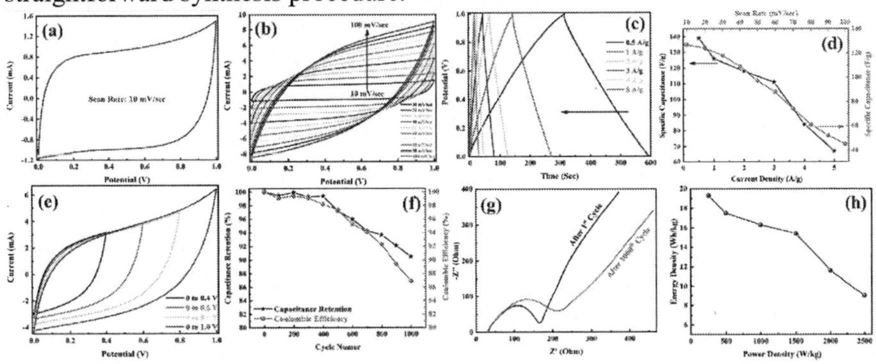

Fig. 3. (a) CV at a scan rate of 10 mV/sec, (d) CVs at different scan rates, (c) GCD profiles at different current densities, and (d) specific capacitance vs. scan rates and current densities of fabricated PCSAC based SC. (e) CVs measured in different OPWs at a scan rate of 50 mV/s, (f) cycling performance measured at a current density of 10 A/g for 1000 GCD cycles, (g) Nyquist plots before and after 1000 GCD cycles, and (h) Ragone plot of the PCSAC based SC.

Conclusions

The chemical activation approach was used to successfully prepare high yield PCSAC for electrochemical energy storage (i.e., SCs) in this study. XRD, Raman spectroscopy, FESEM, EDS, XPS, and BET analyses were used to characterize the produced PCSAC. The amorphous structure and porous character of the produced PCSAC with a BET surface area of ~450 m^2/g were investigated by these results. CV, GCD, and EIS were used to investigate the electrochemical performance of the PCSAC-based SC. At a current density of 0.5 A/g, the PCSAC-based SC with low IR drop displayed excellent EDLC behavior and delivered a good specific capacitance of 139 F/g. Furthermore, the PCSAC-based SC demonstrated good energy and power densities and excellent charging-discharging cycle stability, confirming its superior capacitance efficiency. Therefore, this research is a useful addition due to its simplicity and conciseness on SCs, which can be helpful for researchers working in the field of electrochemical energy storage.

Acknowledgments

The research support provided by the IRC-HES (Project # INHE2105), KFUPM is highly acknowledged.

References

[1] K. Venkat, et al., *J. Energy Storage* **2020**, *31*, 101652.
[2] J. Chmiola, et al., *Science* **2006**, *313*, 1760-1763.
[3] S. S. Shah, et al., *Synth. Met.* **2021**, *277*, 116765.
[4] C. K. Roy, et al., *Chem. Asian J.* **2021**, *16*, 296-308.
[5] N. C. Deb Nath, et al., *ChemistrySelect* **2019**, *4*, 9079-9083.
[6] S. S. Shah, et al., *J. Energy Storage* **2020**, *30*, 101562.
[7] S. S. Shah, et al., *ChemSusChem* **2021**, DOI: https://doi.org/10.1002/cssc.202101282.
[8] N. Nor, et al., *Int. J. Electrochem. Sci.* **2017**, *12*, 2520-2539.
[9] S. S. Shah, et al., *Sci. Rep.* **2021**, *11*, 6945.
[10] D. Zhao, et al., *Adv. Energy Mater.* **2017**, *7*, 1700739.

The Joint Study on GHG Emissions Reduction Technology from Well to Wheel Perspectives in Japanese Automotive Fuel Value-Chain

Kenichirou Saito
ENEOS Research Institute, Inc.

Abstract

JCCP, Saudi Aramco, and ENEOS Research Institute have been jointly studying CO2 reduction technologies of petroleum sector since FY2020. Aiming to explore new collaboration opportunities in the carbon constrained era, we investigated the WtW -GHG emissions and potential reduction technologiesin Japanese automotive fuel value chain. The summary of the interim results will be reported.

We evaluated the GHG emissions of the production and use of the transportation fuel covering the crude production in Saudi Arabia, transportation to Japan, refining, distribution, and vehicle operation. Moreover, we quantitatively assessed the potential GHGs reduction technologies being developed at each stage of the supply chain and explored the pathways to decarbonize the Japanese automobile sector.

We found that a diversity of fuels and powertrain technologies, including advanced combustion engines, low climate impact fuels, vehicle electrification and hydrogen fuel cell, are needed to support an orderly mobility transition. It will require strengthened cooperation between Saudi Arabia and Japan to drive a sustainable Japanese automotive fuel value-chain in the future.

Carbon Recycling Technology Perspectives for CO2 Emission Reduction

Takao Nakagaki
Waseda University

Abstract

For challenge to sustainable energy system accompanied by non-continuous reduction of GHG toward the Net Zero Emissions, recycle use of carbon-containing industrial products essential for our lives is absolutely necessary as well as energy storage feedstocks supporting variable renewable energy as principal power source. Carbon recycling technologies, which are also known and described as carbon dioxide capture, utilization and storage, have been attract attention as a technical option to address the problem. Carbon dioxide-free hydrogen is a key role to convert CO2 into various organic carbon recycling products, such as power to gas, liquid, and chemicals, which reduce new fossil fuel consumption utilizing existing infrastructures effectively. Most of these carbon recycling products can be produced via syngas with conventional catalytic synthesis technologies, and recently, can be produced directly from hydrogen and CO2 via not only innovative catalytic reduction but also electrochemical reduction and synthesis. Capturing CO2 from high-concentration flue gases exhausted by thermal power plants and industrial furnaces has been already commercialized, but DAC technologies, which is expected as one of ultimate carbon sources in the carbon neutral future, are currently under development for further cost and energy reduction. Inorganic carbon recycling products including carbonates using slags and CO2-curing concretes are expected to be widely spread in the near future, which provide environmentally- friendly construction material with a function of long-term CO2 fixation.

Recycle processes from waste plastics to chemical feedstock

Toshiaki Yoshioka
Tohoku University

Abstract

Chemical recycling is attracting attention as a method for achieving both the solution of plastic issues and the carbon cycle. For that purpose, integration of the arterial industry and the venous industry based on the circular economy is necessary. There are great expectations for process development to produce basic chemical feedstock from polymeric materials such as plastics and biomass. In the lecture, I will introduce the effectiveness of applying various arterial industry processes to waste treatment and the technological development for that purpose.

Visible-light driven redox system for CO_2 conversion into valuable organic materials

Yutaka Amao
Osaka City University

Abstract

Carbon recycling technology that can capture and reuse CO_2 for valuable materials and fuels is attracting attention. One of the candidates for carbon recycling technology is the production of low-carbon fuels, hydrogen, CO_2-based alcohol, *etc.* using renewable energy such as solar energy. A solar-driven redox system for the valuable materials production due to building carbon-carbon bonds from CO_2 as the carbon feedstock is also expected as a long-term storage technology for CO_2. In this presentation, some visible light-driven redox systems consisting of a photocatalyst and a biocatalyst for valuable organic materials production due to building carbon-carbon bonds from CO_2 as the carbon feedstock were introduced. In particular, visible light-driven redox system for synthesis of monomer of biodegradable polymer using CO_2 as a raw material is introduced as one of the environmental problem-solving technology.

CO_2 assisted oxidative dehydrogenation of hydrocarbons to produce olefins

Majid L. Balogun[a], Sagir Adamu[a,b], Idris A. Bakare[c], Mohammed S. Ba-Shammakh[a,b],

Mohammad M. Hossain [a,b*]

[a] Department of Chemical Engineering, [b]Center for Refining & Advanced Chemicals, [c]Center for Hydrogen & Energy Storage, King Fahd University of Petroleum & Minerals, Saudi Arabia.

The oxidative dehydrogenation (ODH) of lighter alkanes to olefins are more attractive as compared to the conventional steam cracking/dehydrogenation processes, given the lower energy requirement and possibility of high product selectivity. In ODH, the olefins are obtained by partial oxidation of hydrogen from the alkane chain. In case of long chain alkanes, it also involves sequential cracking of hydrocarbons followed by oxidation of the cracked intermediate into olefins. It is an exothermic reaction - therefore, the energy demand is much lower than that of the cracking or dehydrogenation processes. However, the presence of oxygen may also decrease the selectivity of olefins due to the complete oxidation of both feed alkanes and product olefins into CO_2 and CO. In this regard, the ODH of alkane in the presence of CO_2 is better option than the direct dehydrogenation of alkanes due to the involvement of the reverse water gas shift reaction. It improves the equilibrium conversion by shifting the reaction favorable to the product side and also minimizes the formation coke via the Boudouard reaction. The use of CO_2 as a mild oxidant significantly enhances the selectivity of olefins by suppressing the deep oxidation reactions, although generally, the conversion is lower as compared to commercialized direct dehydrogenation process. In addition to the aforementioned advantages, there are challenges of using CO_2 as an oxidizing agent. CO_2 is a stable molecule and the activation of C=O bond requires high energy. The use of catalyst generally helps to minimize the required energy for bond activation.

The present study investigated the catalytic performance of a novel fluidizable MoO_3 supported on La_2O_3-γAl_2O_3 composite for ODH of propane to propylene with and without CO_2. In general, the propane conversion declined slightly, while propylene selectivity increased with the increasing La_2O_3 content. It was hypothesized that the presence of basic La_2O_3 reduced the acidity and the non-selective sites of γ-Al_2O_3, which were responsible for deep oxidation of propylene and propane. Among the studied catalysts, MoO_3/La_2O_3-$\gamma Al_2O_3$1:2 displayed the highest propylene yield of 28.2 % at 600 °C in the absence of CO_2. However, a superior yield of 35.2% was achieved over the same catalyst by the introduction of CO_2-admixture in the feed at 550 °C.

Key words: Oxidative dehydrogenation; propane; propylene; mild oxidant; CO_2

*Corresponding author: Mohammad M. Hossain
Tel.: +966-13-860-1478; Fax: +966-13-860-4234
E-mail: mhossain@kfupm.edu.sa

Bifunctional tandem catalysts for the one-pass synthesis of lower olefins via CO_2 hydrogenation

Masahira Tada1; Taiga Kinoshita2; Kazu Ochiai2; Chokkalingam, Anand3; Hu, Peidong3,4; Noriko Yamauchi1; Yoshio Kobayashi1; Kenta Iyoki4

1 Ibaraki University Graduate School of Science and Engineering (Engineering Field), Material Science and Engineering Field
2 Ibaraki University Graduate School of Science and Engineering (Engineering Field), Quantum Line Science
3 Graduate School of Engineering, The University of Tokyo, General Research Institute
4 Department of Chemical Systems Engineering, Graduate School of Engineering, The University of Tokyo

Abstract

The development of methodologies for CO_2 utilization is in high demand worldwide. Here, we propose bifunctional tandem catalysts of $ZnZrO_x$ (for CO_2-to-methanol hydrogenation) and a series of solid acid catalysts (for subsequent methanol conversion to light olefins). As solid acid catalysts, we used zeolites and silicoaluminophosphates with different topologies, MOR, FER, MFI, *BEA, CHA, and ERI. The use of tandem catalysts realized a one-pass synthesis of lower olefins, while no hydrocarbons were obtained using $ZnZrO_x$ only. The pore sizes and channel dimensionality of zeolites influence the selectivity of products; zeolites with small pores, such as MOR, SAPO-34 and ERI, are promising, while zeolites with larger pores, such as MFI, produce heavier hydrocarbons. The results provide new insight into the design of innovative catalysts for CO_2 utilization.

Catalytic light alkanes conversion: Is it contributable on the way to carbon neutral realization?

Wataru Ueda
Kanagawa University

Abstract

Natural gas utilization for energy and chemicals production in one hand and suppression of CO_2 emission toward NetZero in other hand are obviously contradictory in a sense of chemical process. However, since main energy and chemicals production technologies seem flowing to the direction of gas utilization basis after passing the coal and petroleum utilization ages, harmonic advancement of natural gas utilization technologies and of CO_2-relating ones at the same time is indispensable for sustaining our human society as well as global environment in coming years. In order to realize and to keep such harmonicity, highly advanced catalytic technologies must be a key for both the utilizations and more specifically selective light alkane catalytic oxidation to chemicals will be more important than before because the process is highly energy productive and desirably low CO_2 emission if catalysts are developed extremely selective for products without accompanying CO_2 formation. The presentation will introduce recent research activity on catalytic methane oxidation and ethane oxidation to ethylene and will discuss how this catalytic reaction can contribute to the way to carbon neutral realization.

Development of versatile spheroidal $MgCl_2$ polyolefin catalyst support and related innovation and spin-off challenges

Muhammad Atiqullah
E-mail: matiq@kfupm.edu.sa
Center for Refining & Advanced Chemicals

EXTENDED ABSTRACT

Twenty Saudi petrochemical-polymer industries produce about 16.25 mmtpa polyolefin. This huge production volume ranks Saudi Arabia as the first and the 4th largest polyolefin (PO) producer in the Middle East and across the world, respectively. Additionally, the country has a vast number of polyolefin compounding and plastics processing industries. The overall Saudi petrochemical and polymer business continues to remain progressing and profiting because of producing particularly polyolefins. Therefore, from a total perspective and on principle, Saudi Arabia requires to be an international polyolefin education, research, training, and business hub and leader. Figure 1 depicts the Center for Refining & Advanced Chemicals (CRAC) modest effort, particularly in the PO catalyst research arena.

Transition metal catalytic processes produce polyolefins. This catalyst field remains outstandingly prolific and active over 70 years since the pioneering and seminal discoveries of Nobel laureates Karl Ziegler and Giulio Natta in early 1950s. The impact of these discoveries on polyolefin education, research, training, industry, and worldwide business is so deep and comprehensive. PO catalyst research is an unending journey with unreachable destination. Many later generation (metallocene and post-metallocene) novel laboratory-scale precatalysts are not yet fully exploited because of the challenges facing the development of supported PO catalysts. Such catalysts are greatly required by industrial slurry and fluid bed gas phase olefin polymerization processes. Precatalyst immobilization and polymerization rate limitation cause dormant and/or partially accessible catalytic sites. This situation is technically unacceptable while the following is well established. Silica, the customary non-Lewis acid support, has various surface functionalities (isolated, geminal, and vicinal hydroxyl groups, as well as constrained siloxane) which originate from the synthesis process. Moreover, it serves, upon impregnation, as a giant ligand to the precatalyst. All these add, as a result, significant heterogeneity to the supported catalyst and poses large steric hindrance to the incoming olefin. Consequently, the catalyst activity drastically drops, which poses an enormous technical challenge. Can this challenge be overcome? The answer is apparently **yes**.

To address the above challenge, first, this talk will review the current PO catalyst subject from technical perspective as well as market-insight. As per the catalyst active species and mechanism of olefin polymerization, a Lewis acid spheroidal and disintegrable support is proposed to address this challenge. Note that seven major process technologies and suppliers that make different grades of silica are worldwide available. The Lewis acid $MgCl_2$ support has been predicted to hold a great potential. However, process technologies and suppliers, marketing various grades of commercial $MgCl_2$ supports (having varying properties), are not comparably available. This motivated us to conduct some modest research in this area.

Second, our preliminary success as well as failures, which originate from the above research, will be summarized. Figure 2 illustrates the emulsifier-free support synthesis work while Figure 3 shows several successful trial $MgCl_2$•EtOH adduct morphology. But we faced the following problem. The process did not turn out to be robust with respect to change in process variables, particularly introduction of structure-modifying additives. Nevertheless, our learning and insight set new future research direction, which we hope will make a significant difference in the subject support manufacture. The development and commercialization of a process that makes various grades of spheroidal amorphous to crystalline $MgCl_2$ supports will revolutionize the PO catalyst applications and processes.

Third, the possibility of establishing an independent spin-off company in this area will be evaluated while King Fahd University of Petroleum & Minerals currently undergoes a paradigm shift in research. Here, innovation with impactful industrial values comes into play. Hence, the innovation diffusion that pertains to—research and researchers, R&D management and managers, and potential investment and investors—will be highlighted. See Figure 4. *The critical need is to identify and understand why an innovation will fail to be accepted and marketed.* Finally, the importance of conducive sociology (environment), psychology (mental state), and mindset (preparedness to make and accept changes) that precede and dominate over innovation and technology will be consolidated.

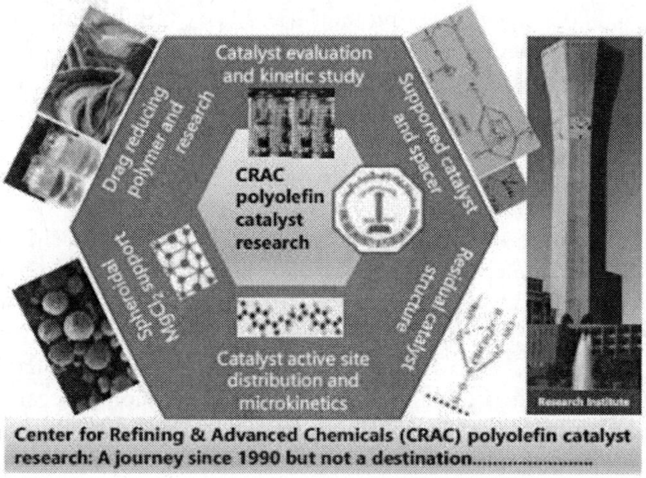

Figure 1. Polyolefin catalyst research at the Center for Refining & Advanced Chemicals, RI, KFUPM.

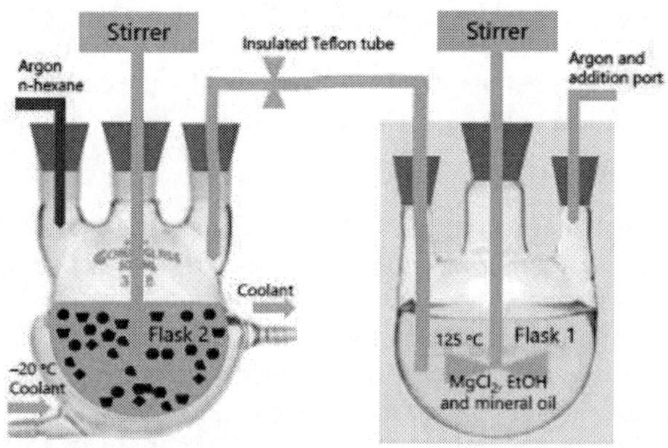

Figure 2. Emulsion-free two-vessel $MgCl_2$•EtOH support synthesis set-up.

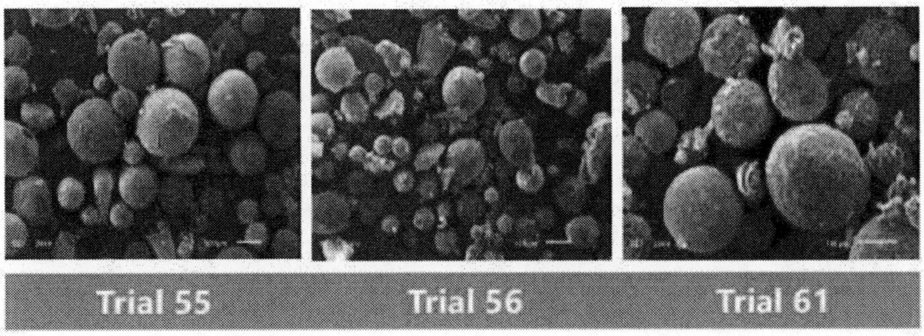

Figure 3. Formation of spheroidal $MgCl_2$•EtOH particle morphology with preliminary success.

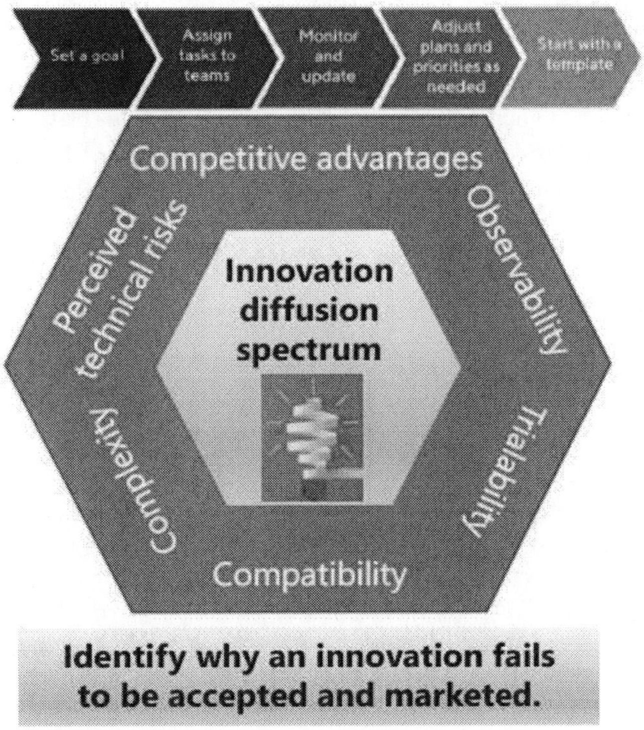

Figure 4. Application of innovation diffusion concept to product and technology development and industry investment.

Advanced FCC catalyst design for LPG production

Yusuke Takamiya
JGC C&C Ltd., Japan

Abstract

Role of fluid catalytic cracking (FCC) process has been changing in this decade. Previously, the FCC process played an important role in gasoline production. However, the demand of gasoline decreased because ecological car such as electric vehicles spread. On the other hand, the demand of chemicals such as propylene has been increasing in this decade. Thus, FCC units are required to change the target product from fuel to chemicals. In terms of FCC catalyst design for increasing propylene yield, it is necessary to suppress hydrogen transfer reaction by applying low RE / zeolite ratio or USY with low unit cell size. The FCC catalyst, based on such the design, shows much olefins yield in gasoline and they would be cracked to light olefins on the added ZSM-5 additive.

We have challenged to satisfy the demand by means of changing properties of USY zeolite included in FCC catalyst and have found a promising FCC catalyst for such demand. The FCC catalyst with Al ion exchanged USY has indicated higher LPG yield than a conventional FCC catalyst with La ion exchanged USY. This study provides an advanced FCC catalyst design for LPG production. The effect of Al ion exchanged USY on VGO cracking activity and catalyst properties will be reported in the presentation.

Hydrodearylation: A new process to enhance BTX yields in an aromatics recovery complex

Robert P. Hodgkins*[a], Omer R. Koseoglu[a] and Vinod Ramaseshan[b]

[a]Refining and Upgrading Division, Research & Development Center, Saudi Aramco, Dhahran, 31311 Saudi Arabia
[b]Downstream Process Engineering Division, Process & Control Systems Department, Saudi Aramco, Dhahran, 31311 Saudi Arabia

ABSTRACT

With the increasing demand for petrochemicals set to continue over the next decades realizing maximum value requires integrating fully across the hydrocarbon value chain. The rapid growth in chemicals demand provides a real opportunity to pursue new refining processes to integrate petrochemicals into an expanding global refining and marketing system. Research within the Research & Development Center at Saudi Aramco has led to the development of a new refining process to upgrade a low-value aromatic reject stream, derived from an aromatics recovery complex, to BTX and petrochemical feedstock. The hydrodearylation process specifically creates value from the mono-aromatic loss realized at the clay treater within an aromatics complex, *via* the formation of heavy non-condensed di-aromatics, by selectively targeting the cracking of the bridging alkyl group of the non-condensed di-aromatics to form mono-aromatics. Typically, these heavy non-condensed di-aromatics are used as low-value fuel oil blending components. The hydrodearylation process converts this fuel oil blending component directly into BTX and petrochemical feedstock.

CATALYSIS BY DESIGN: SYNTHESIS OF WELL-DEFINED TI AND ZR SURFACE COMPLEXES ON ULTRA STABLE ZEOLITE (Y) FOR REFINERY PROCESS

Manoja K. Samantaray, Rajesh Parsapur, Amol Hengne, Swechchha Pandey, Anissa Bendjeriou-Sedjerari, Kuo-Wei Huang*, Magnus A. Rueping*

Kaust Catalysis Center, King Abdullah University of Science and Technology, Thuwal, 23955-6900, KSA.

Corresponding author: manoja.samantaray@kaust.edu.sa

Abstract

A series of well-defined ultra-stable zeolite (Y) supported Ti and Zr surface complexes were synthesized and fully characterized using modern NMR techniques. The corresponding supported metal hydride species were also synthesized and fully characterized using FT-IR, ICP, and NMR techniques. These supported catalysts were tested for conversion of wax to valuable chemicals.

Keywords: Hydrogenolysis, Organometallic complex, Ultra stable zeolite (USY).

Introduction

Heterogeneous catalysis, a field important industrially and scientifically, is increasingly seeking and refining strategies to render itself more predictable. The main challenge is the understanding of the nature and the population of catalytically active sites. Their number is generally very low, their "acid strengths" or "redox properties" is not homogeneous, and the material may display related yet inactive sites on the same material. In many heterogeneous catalysts, the discovery and understanding of a structure-activity relationship is challenging. One possible solution passes by generating single-site catalysts in which most of, if not all, the sites are structurally identical. Within this context and using the right tool, the catalyst structure can be well defined that is elucidated to reach a molecular understanding. It is then feasible to understand structure-activity relationships and to develop predictable heterogeneous catalysis. "*Single-site well defined heterogeneous catalysts can be prepared using Sol-Gel Chemistry and Surface Organometallic Chemistry*."

Zeolites, specifically, have strong Bronsted acidity, high surface area, high (hydro)thermal stability and shape selectivity, making them the "perfect" catalysts for a panel of reactions in the petroleum industry. However, they feature a major drawback in terms of access and diffusion limitations. To overcome these limitations, recently, we developed a new methodology to synthesize highly porous USY zeolites and grafted tetra neopentyl Zr and Ti on the surface (Scheme 1 and Figure 1).

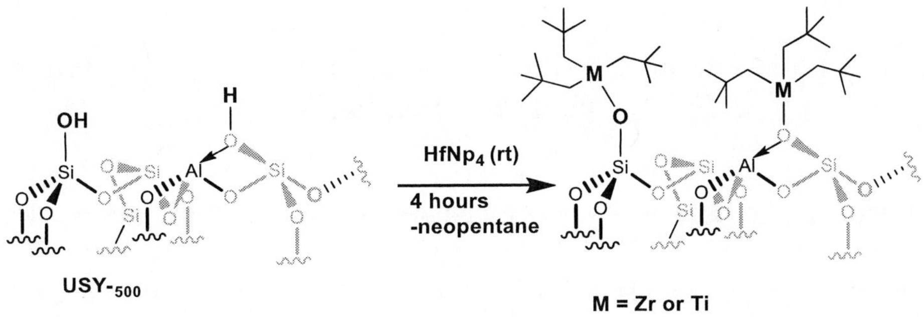

Scheme 1. Synthesis of USY supported Zr and Ti complex presented here

Conclusions

In conclusion, we synthesized well-defined Zr/Ti surface complexes on USY. These complexes are fully characterized using modern characterization techniques and tested the catalysts for hydrogenolysis of wax into valuable chemicals.

Enhanced Light Olefins Production using Modified MFI Catalyst in n-Pentane Cracking

Ziyauddin S. Qureshi, M. A. Bari Siddiqui, Abdullah M. Aitani, Hassan Alasiri

Center for Refining & Advanced Chemicals, King Fahd University of Petroleum & Minerals, Dhahran 31261, Saudi Arabia

Abstract

The increased demand for light olefins in most polymer and petrochemical industries sparked interest in the development of novel production techniques based on alkanes as the starting raw material. These olefins (ethylene and propylene) are regarded as the petrochemical industry's backbone and generally obtained from the steam pyrolysis/cracking of C_2-C_4 alkanes, naphtha, light diesel, as well as heavier hydrocarbons at temperature ranging from 800-1000 °C.[1] High process temperature is the major constraint in steam cracking, accounting for 40% of total energy required with in petrochemical industry.

In this presentation, we will discuss about the catalytic cracking of n-pentane over a series of MFI zeolites with SiO_2/Al_2O_3 molar ratio (30-1500), different crystal size (nano and large crystal) and surface modified zeolite (ammonia and phosphorous). Our findings reveal dramatic differences in catalyst performance over the phosphorous modified MFI catalyst, with improved propylene selectivity with P/E = ~1.3 and high yield of light olefins (47.2%) at 650 °C. Collectively, this work provides new insights into the design of phosphorous modified MFI zeolite catalysts for light naphtha upgrading.

Reference:

1. Amghizar, I., Vandewalle, L.A., Van Geem, K.M., and Marin, G.B. (2017). New Trends in Olefin Production. Engineering, *3*, 171−178.

Novel route for on-purpose production of butadiene

Gazali Tanimu, Abdullah M. Aitani, Sachio Asaoka, Hassan Alasiri

Center for Refining and Advanced Chemicals, Research Institute, King Fahd University of Petroleum and Minerals, Dhahran 31261, Saudi Arabia.
Email address: gazali.tanimu@kfupm.edu.sa

Abstract

1,3 Butadiene (BD) is an important raw material in the chemical industry utilized in the manufacturing of synthetic rubber and automotive tires. Conventionally, BD is a produced as by-product within the C4 fraction of the steam naphtha cracking processes. Currently, petrochemical industries are using lighter feedstocks to produce ethylene and propylene, which means less amount of BD is co-produced. An alternative route for on-purpose BD production is the oxidative dehydrogenation (ODH) reaction aimed at supplementing the existing supply methods, in order to meet the increasing demand. The major challenge with the ODH reaction is the selectivity control resulting from competitive series reactions to deep oxidation products. An efficient catalyst design and choice of reaction conditions can improve the selectivity to the desired BD. Catalyst systems comprising of Ni and Bi metal oxides supported on Al_2O_3, SiO_2, and ZrO_2, have shown promising selectivity to BD. In line with those catalyst systems, Ni-Bi-O/mesoporous silica (MCM-41, SiO_2 foam, and SBA-15) catalysts containing 10-20wt% Ni and 10-30wt% Bi, were utilized for oxidative dehydrogenation of n-butane to BD. SBA-15 supported catalyst showed a better performance compared to the other mesoporous silica supported catalysts. The active Ni and Bi metal oxides loading were varied. It became clear that, the Bi_2SiO_5 and beta-Bi_2O_3 phases changed with the Ni/Bi loading amount and ratio due to changing template effect of Ni to Bi oxide as evident from XRD measurement. The degree of bismuth oxide phases formation reflected in the selectivity through redox cycle stabilization and acid/base characters' moderation, as evident from H_2-TPR and TPD measurements, respectively. 14 wt % Ni 10 wt % Bi-O/SBA-15 catalyst, containing moderate loadings of both of Ni and Bi, showed a clear superiority in the performance (BD selectivity of 53 % at 27.7 % n-butane conversion) compared to the other catalysts. This catalyst also showed superior performance for the conversion of n-butenes.

Keywords: oxidative dehydrogenation; *n*-butane; butadiene; mesoporous silica; Ni oxide.

REFERENCES

[1] B. R. Jermy, B. P. Ajayi, B. A. Abussaud, S. Asaoka, S. Al-Khattaf, J. Mol. Catal. A Chem 400 (2015) 121-131
[2] G. Tanimu, B. R. Jermy, S. Asaoka, S. Al-Khattaf, J. Ind. Eng. Chem 45 (2017) 111-120
[3] G. Tanimu, S. Asaoka, S. Al-Khattaf, Mol. Catal. 438 (2017) 245-255
[4] G. Tanimu, A. Palani, S. Asaoka, S. Al-Khattaf, Catal. Today 324 (2019) 97-105
[5] G. Tanimu, A.M. Aitani, S. Asaoka, H. Alasiri, Mol. Catal. 488 (2020) 110893

Controlled autoxidation of hydrocarbon to produce value-added chemicals

Muhammad N. Siddiquee* [a] [b], Mohammad M. Hossain a] [b]

a Department of Chemical Engineering
b Interdisciplinary Research Center for Refining & Advanced Chemicals (IRC-RAC), King Fahd University of Petroleum & Minerals (KFUPM), Dhahran 31261, KSA

*E-mail: muhammad.siddiquee@kfupm.edu.sa

Extended Abstract

Liquid-phase oxidation of hydrocarbons with air or molecular oxygen (also known as autoxidation) is an important process in petrochemical and pharmaceutical industry [1-3]. It is a cheap and sustainable technique that follows a complex free radical mechanism involving initiation, propagation, and termination. It produces different intermediate products as well as final products (known as oxygenates having oxygen functional group) namely peroxides, alcohols, ketones, ester and acids following a reaction network (Figure 1). The key challenge of this free radical process is to control the product selectivity. In industry, low-conversion is usually practiced to control the product selectivity.

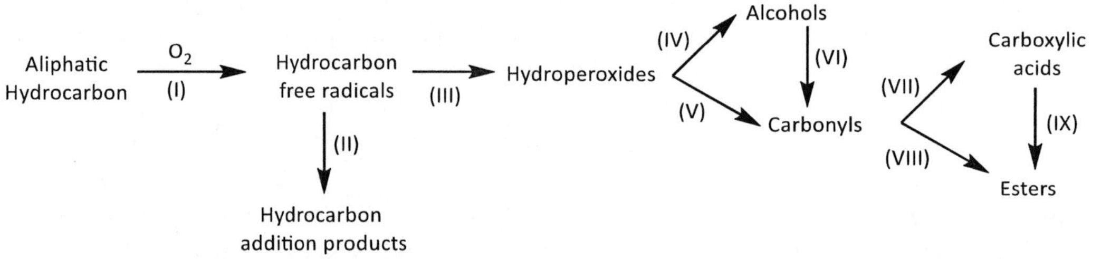

Figure 1. Overall reaction network of the aliphatic hydrocarbon autoxidation [1, 3].

Different steps proceed at different rates- initiation (Step I) is slow, but propagation and termination steps (Steps II- IX) are fast. For instance, the rate of tetralin initiation ($RH + O_2 + RH \rightarrow 2R\cdot + H_2O_2$) with molecular oxygen at 150 °C is 7.1×10^{-8} L^2/mol^2. s, whereas the rate propagation ($R\cdot + O_2 \rightarrow 2ROO\cdot$) is 6.8×10^7 L/mol. s, and the rate of termination ($2ROO\cdot \rightarrow 2ROH + R_1COR_2 + O_2$) is 2.6×10^7 L/mol. s, and hydroperoxide decomposition rate ($ROOH \rightarrow RO\cdot + HO\cdot$) is 2.3×10^{-4} s^{-1} [5]. Oxygenates that produced are surface actives and more reactive than the parents hydrocarbons. Different reaction steps at different rates make it complex to control the product selectivity. Moreover, hydrocarbon and oxidants are present in liquid and gas phase, respectively. The propagation step is considered as oxygen transfer limited as the reaction between alkyl radical and oxygen is very fast [4]. The free radicals are highly reactive and depending on the oxygen availability they can form addition product (Step II). Therefore, oxygen availability in the liquid phase from the gas phase is an important factor to control the product selectivity. Temperature, type of hydrocarbon, mixing and the residence time are also the influencing factors that affect the conversion and the product selectivity.

With the advances of the reactor technology, it is possible to understand and control the different steps of the free radical processes to manipulate the conversion and the product selectivity. For example, initiation period of tetralin oxidation at 150 °C in a semi-batch reactor is about 14 minutes (Figure 2). During this period there was insignificant free radical formation and oxygen consumption. Experiment performed in microfluidic reactor showed induction period can be significantly reduced by manipulating gas-liquid interfacial area (*a*) ensuring very high oxygen availability [5, 6].

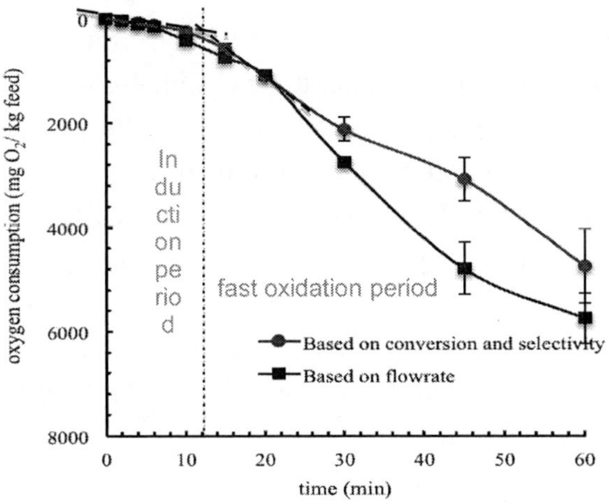

Figure 2. Oxygen consumption over time during the tetralin oxidation of tetralin at 150 °C in a Semibatch reactor [triplicate results: (●) based on conversion and product selectivity (water generation was not considered)) and (■) based on oxygen flowrates] [5].

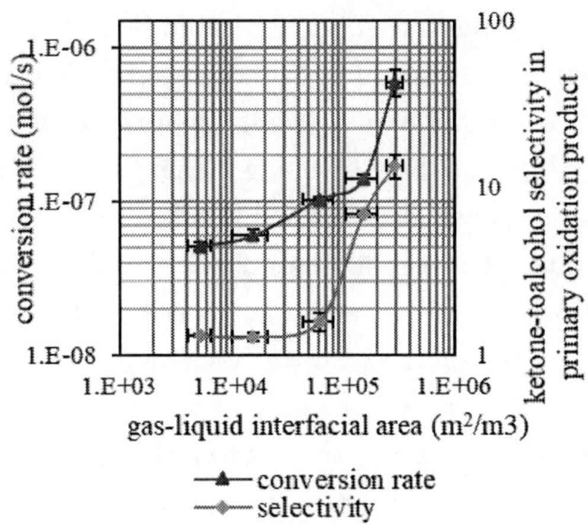

Figure 3: Tetralin oxidation with oxygen at 150 °C and 90 kPa gauge pressure in a microfluidic reactor – typical residence time 2-3 minutes[5].

Tetralin oxidation study with molecular oxygen in a 1 mL rectangular microfluidic reactor maintaining slug flow showed that conversion rate and product selectivity were significantly increased with oxygen availability (gas liquid interfacial area, *a*) (**Figure 3**) [5]. It was found that ketone-to-alcohol formation in primary oxidation product was enhanced from 1:1 to 14:1 by manipulating the oxygen availability at near constant conversion (**Figure 3**). In compare to batch and semi-batch reactors**,** this selectivity change was significantly higher. The main reason of higher ketone selectivity would be the following reactions:

$$RO\cdot + O_2 \rightarrow R{=}O + HOO\cdot$$

$$ROH + 0.5\ O_2 \rightarrow R{=}O + H_2O$$

Indan oxidation study with molecular oxygen in a 62.5 μL half-elliptical microfluidic reactor (Reactor A) and a 1000 μL rectangular microfluidic reactor (Reactor B) confirmed that gas-liquid interfacial area (*a*) greatly influenced conversion rather than the selectivity [6]. The indan conversion in Reactor A and Reactor B were 28.0 wt/wt% and 11.5 wt/wt%, respectively, at the same experimental conditions of 150 °C, 300 kPa pressure and indan injection rate was 2 μL/min. The gas-liquid interfacial area of Reactor A was 7.8×10^5 m^2/m^3 and main contributor of such higher *a* was the smaller reactor dimension. The film surrounding the gas bubble in Taylor flow was the main contributor of gas-liquid interfacial area (*a*) of 1.4×10^5 m^2/m^3 for Reactor B (1000 μL, rectangular). The reactor size and shape influence the oxygen availability and hence, the conversion and selectivity differently. It showed how engineering could be used to control the chemistry. The understanding from the study could be used in reactor design and operation of liquid phase oxidation of hydrocarbons to produce value-added products and to use as fine chemicals and pharmaceuticals.

Kingdom of Saudi Arabia (KSA) is one of the leading crude oil producing country in the world. It produced huge amount light alkanes and heavy petroleum during upgrading/refining process of crude oils. Light naphtha fractions has lower economic value. It is possible to functionalize them by using a cheap and sustainable oxidising agent- air to produce alcohol and ketone. These alcohols and ketones have higher market value. One of the prominent way to produce alcohol and ketone is the catalytic selective oxidation. However, traditional catalytic route is expensive, generates waste, and requires catalyst recovery. It is important to develop an alternative and economically viable process to functionalize light hydrocarbons to produce value added chemicals from light naphtha fractions. The concept can also be applied to transform lowest value heavy fraction to produce high value products. Overall, the controlled autoxidation of hydrocarbon is a promising way to produce value-added products.

Acknowledgements

The author(s) would like to acknowledge the support provided by the Interdisciplinary Research Center for Refining & Advanced Chemicals (IRC-RAC) at King Fahd University of Petroleum & Minerals (KFUPM) for funding this work through project No. INRC2013.

References

1. N. M. Emanuel, E. T. Denisov and Z. K. Maizus, *Liquid-phase oxidation of hydrocarbons*; Plenum Press: New York, 1967.
2. A. K. Suresh, M. M. Sharma and T. Sridhar, *Ind. Eng. Chem. Res*., 2000, **39**, 3958–3997.
3. A. de Klerk, *Fischer-Tropsch Refining*, 1st ed.; Wiley VCH: Weinheim, 2011.
4. A. Günther, SA. Khan, M. Thalmann, F. Trachsel, K. Jensen, "Transport and reaction in microscale segmented gas–liquid flow", Lab *Chip*, 4, 278–286, 2004.
5. M. N. Siddiquee, A. de Klerk and N. Nazemifard, *React. Chem. Eng.*, 2016, **1**, 418–435.
6. M. N. Siddiquee, Y. Wu, A. de Klerk and N. Nazemifard, *J. Flow. Chem.,* 2020, 10, 647-660.

Impact of Hydrogen Transfer Reactions in Catalytic Processes on Olefins Yield

Shatha A. Alabbad
Saudi Aramco R&DC

Abstract

Fundamental understanding of the kinetics and mechanisms of essential reactions in catalytic cracking of crude oil is an imperative task in academia and industry. The excessive formation of paraffins in cracking processes, for instance, is one of the puzzles to resolve to keep up with the spiking demand of high-value chemicals while reducing capital and operational carbon. This study reviews the effects of temperature (450-525°C), contact time (3-7s), and zeolite structure on the overall conversion of a model component and its selectivity towards C_2-C_4 olefins in the riser simulator. Cyclohexane was selected to enable the quantification of contribution of hydrogen transfer reactions to paraffins, aromatics and coke formation. Thus far, surprisingly, initial findings in analyzing results at a contact time of 7 s show a paraffins-to-olefins ratio of 3.2 at 450°C which decreases at elevated temperatures reaching nearly 1.8 at 525°C. At reaction conditions, approximately, 52-70 mol% of formed olefins are re-consumed in hydrogen transfer reactions to byproducts over the temperature range where naphthene donates hydrogen to an olefin and becomes aromatic and the olefin is hydrogenated into a paraffin. The data indicates that different reaction pathways can take place at different temperatures. In addition, the investigation results suggest that steaming the catalyst reduces hydrogen transfer reactions increasing olefins yield. The impact of catalyst modification and reaction conditions on the rate of hydrogen transfer reactions will be further illustrated in this presentation.

Maximizing Olefins from Crude – UOP Integrated Olefins Suite

Priyesh Jani
Honeywell UOP

Abstract

Steam cracking technology is used to produce over 97% of the world's ethylene, which is forecasted to grow at a rate of about 3.5% through 2030. Ethane and crude derived naphtha are the most common feedstocks with each contributing approximately 40% towards global ethylene production. Naphtha tends to be the more common feedstock in Asia, Europe and other regions due to its availability and ease of transport. In addition to higher production cost, naphtha crackers, however, suffer from relatively low selectivity to ethylene (typically 30%-35%) with significant co-production of lower value by-products. To help naphtha cracking operators improve competitiveness, UOP has developed the ***Integrated Olefin Suite (UOP-IOS).*** The UOP-IOS is a collection of proven technologies that synergistically integrate with steam cracking units to tune the overall product slate to match the operator's strategic objectives and improve profitability. Case examples will be shared demonstrating some of the capabilities of the UOP Integrated Olefin Suite.

Technology in Petroleum Refining & Petrochemicals

Innovation towards Circular Carbon Economy

ONLINE POSTERS

Day Two: Tuesday, December 14, 2021 10:30-11:30

POSTER SESSION — *Moderator: Dr. Nabeel S. Abo-Ghander*
KFUPM CHE

Each poster is 7 minutes — *Program as of Nov. 22, 2021*

Rapid synthesis of Hf-Beta zeolite for upgrading bio-based furan derivatives

Ryoichi Otomo1, Taichi Nakamura2, and Yuichi Kamiya2
1 Faculty of Environmental Earth Science, and 2 Graduate School of Environmental Science, Hokkaido University, Japan

ABSTRACT

Hf-Beta is a promisingly active Lewis acid catalyst for Meerwein-Ponndorf-Verley (MPV) reduction and other important organic reactions. However, the conventional hydrothermal synthesis of Hf-Beta inevitably requires the troublesome procedure for hydrolysis of tetraalkoxysilane and the long period to complete crystallization. In the present study, we applied a synthetic approach of reducing water content of precursor gel and succeeded in considerably shortening the period for crystallization of Hf-Beta using fumed silica as silicon source. By using precursor gel with H_2O/SiO_2 = 1.2 – 7.0, effect of water content of precursor gel on the crystallization of Hf-Beta, the incorporation of Hf into the zeolite framework, and the catalytic performance were thoroughly investigated. Low water content was favorable for accelerating the crystallization, but unfavorable for the incorporation of Hf into the zeolite framework, resulting in poor catalytic activity. With the assistance of seed crystal, Hf-Beta with a relatively large amount of framework Hf was obtained from the precursor gel with H_2O/SiO_2 = 5.9 in 24 h. Hf-Beta synthesized in this way showed higher catalytic activity for MPV reduction than Hf-Beta as well as Zr-, and Sn-Beta synthesized by the conventional hydrothermal method.

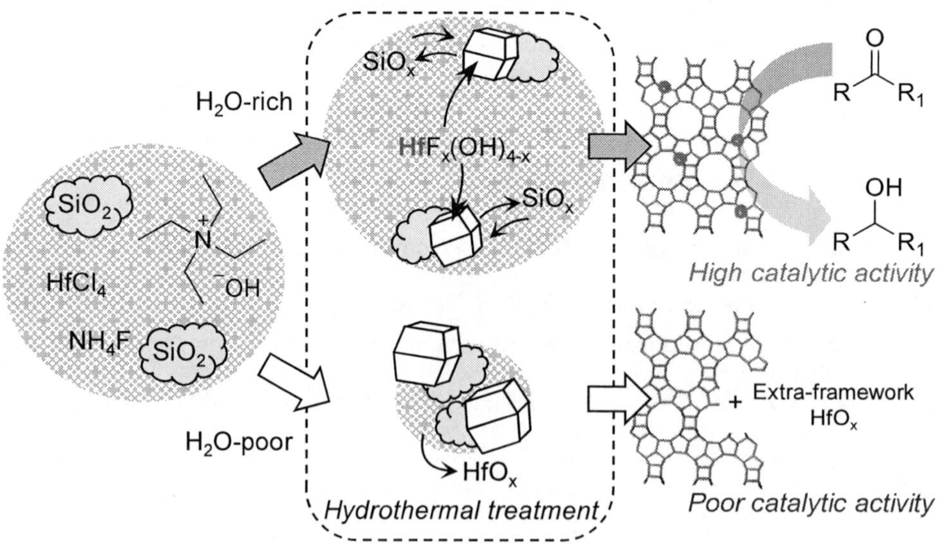

Cracking of botryococcene as chemical utilization of algae

Ryota Miyazaki*1・Kotaro Takeyasu*1・Tadahiro Fujitani*2・Junji Nakamura*1

(University of Tsukuba*1・National Institute of Advanced Industrial Science and Technology2

ABSTRACT

- Botryococcene brownie, a type of microalgae accumulates high-value hydrocarbon oil botryococcene (C34H58) in the extracellular matrix at a high concentration, so its utilization as a non-oil resource is attracting attention
- The purpose of this study is to convert botryococcene into useful hydrocarbons by cracking at atmospheric pressure.
- We optimized the reaction conditions using a flow reactor and evaluated the decomposition reaction and catalytic cracking of botryococcene.
- In pyrolysis, the decomposition progresses gradually as the temperature increases, and the ratio of C1-5 increases.
- In catalytic reaction, decomposition was accelerated by TiO2
- It was suggested that the decomposition temperature and products could be controlled by using catalysts.

Stable Heterogenized Pd-NHC catalysts for Carbonylation Reactions

Waseem Mansour[a], Bassam El Ali[a,b], Mohammed Fettouhi[a,b], and Wissam Iali[a,b]

a) King Fahd University of Petroleum& Minerals, Chemistry Department, Dhahran
b) King Fahd University of Petroleum& Minerals, Center for Refining & Advanced Chemieals, Dhahran

Abstract

Exceptionally stable, active and recoverable heterogenous catalysts were synthesized by supporting propylene-bridged bis(N-heterocyclic carbene)palladium(II) complexes on Merrifield's resin via covalently anchored bond. The new heterogeneous catalysts and their precursors-ligands were fully characterized using various physical and spectroscopic techniques like FT-IR spectrum, ^{1}H NMR, solid and liquid ^{13}C NMR, SEM spectroscopy, powder XRD, X-ray photoelectron spectroscopy (XPS) spectrum, TGA spectra and CP/MASS. The catalytic efficiency and recyclability of the new anchored heterogeneous catalysts were investigated in carbonylative Sonogashira coupling reactions of aryl iodides with aryl and alkyl alkynes. *This work* comprises a comparative study between the catalytic activity of the heterogeneous catalysts and the equivalent homogenous propylene-bridged-$(NHC)_2$Pd complexes.

Ultra-Sensitive-Fast NMR Characterization of Sulfur-Heterocyclic Compounds Found in Petroleum

Wissam Iali[a,b], Bassam El Ali[a,b], Mohammed Fettouhi[a,b], and Waseem Mansour[a]

a) King Fahd University of Petroleum& Minerals, Chemistry Department, Dhahran
b) King Fahd University of Petroleum& Minerals, Center for Refining & Advanced Chemicals, Dhahran

Abstract

Signal Amplification by Reversible Exchange (SABRE) is a new method of hyperpolarization was used to enhance the NMR detection of sulfur-heterocycles, which are typical compounds found in petroleum and refined petroleum products.
The hyperpolarization of S-heterocycles compounds by SABRE (can be done in seconds) offers potential advantages of providing structural information and quantity percentage about sulfur-containing contaminants in petroleum, thereby informing petroleum purification and refining to minimize sulfur content in refined products such as gasoline. Furthermore, this sensitivity gains in NMR spectroscopy given by SABRE-hyperpolarization will allow to perform structural analysis of petroleum products using low-cost, low-magnetic-field (ca. 1 Tesla magnet) high-resolution NMR spectrometers which ideally preferred by the industries in the field.

Highly Efficient NHC-iridium(I) Catalyst for Green Oxidative Coupling Reaction of Thiols

Wissam Iali[a,b], Rami Suleiman[c], Bassam El Ali[a,b], and Mohammed Fettouhi[a,b]

a) King Fahd University of Petroleum& Minerals, Chemistry Department, Dhahran
b) King Fahd University of Petroleum& Minerals, Center for Refining & Advanced Chemicals, Dhahran
c) King Fahd University of Petroleum& Minerals, Interdisciplinary Research Center for Advanced Materials, Dhahran

Abstract:

The N-heterocyclic carbene (NHC) iridium(I) complex [IrCl(COD)(Imes)] [Imes=1,3-bis(2,4,6-trimethylphenyl) imidazole-2-ylidene] howed an excellent catalytic activity in the reactions of the aerobic oxidative coupling of thiols into disulfides. The use of oxygen of air as a green oxidant allows the thiols to be quantitatively converted into the corresponding disulfides under mild conditions and in a short reaction time. The replacement of chloride ligand on the iridium by the strong-donating thiolate ligand yielded an electron-rich and sensitive-to-oxygen active species [Ir(S-R)(COD)(Imes)] at room light, and subsequently promoting the catalytic oxidative-coupling reactions. This active species was successfully isolated and characterized by X-ray diffraction and found also to be essential in revealing a plausible mechanism for this Ir-catalyzed transformation. The DFT-calculated free energy profile of the proposed mechanism of our developed catalyst system can reasonably account for the key role of air in pursuing the catalytic reaction. Overall, the obtained computational and the experimental results of this study are in good agreement.

Development of a shape-stabilized phase change material utilizing natural and industrial byproducts for thermal energy storage in buildings

KHALED OWN MOHAISEN, g201406580@kfupm.edu.sa

Md Hasan Zahir, hzahir@kfupm.edu.sa

KFUPM Interdisciplinary Research Center in Renewable Energy and Power

ABSTRACT

A comprehensive study was conducted to develop and utilize a novel shape-stabilized phase change material utilizing two abundantly available and low-cost natural materials, namely scoria and expanded perlite and an industrial byproduct, heavy oil ash, in combination with polyethylene glycol. The thermal and energy storage characteristics of the composite materials were evaluated with the aim of using them to conserve energy in the domestic facilities. The results of differential scanning calorimetry showed that expanded perlite composite has the highest melting and solidification latent heat values, 150.7 J/g and 134.6 J/g, respectively, compared to scoria and oil ash composite. However, expanded perlite composite has lower thermal conductivity compared to other composites. Consequently, a novel system incorporating carbon nano tubes (0.5 wt.% and 1 wt.%) in the expanded perlite composite was developed to improve its thermal conductivity. The thermal conductivity (0.453 W/m.K) of the new system with 0.5% carbon nano tubes is remarkably more than that of commonly used phase change materials. Further, the developed PCM with 0.5% carbon nano tubes can transform sunlight into thermal energy with a solar-to-thermal energy conversion efficiency of 59.4% and it has a thermal conductivity that is 97% more than that of polyethylene glycol alone. Besides, the newly developed PCM also shows excellent energy storage and release performance. All these favorable characteristics indicate that the developed phase change material can be beneficially utilized in thermal storage systems.

Fuel Design Using Genetic Algorithms and Artificial Neural Networks

Faisal Alboqami 1, Vincent van Oudenhoven 2, Dr. Abdul Gani Abdul Jameel 1

1 Department of Chemical Engineering, King Fahd University of Petroleum and Minerals, Dharan, Saudi Arabia
2 Electrical Engineering, Mathematics and Computer Science, Delft University of Technology, Delft, Netherlands

ABSTRACT

The advancement during the era of the fourth industrial revolution lies in minimizing or eliminating human interference. This broadly applicable concept has found its course into the oil and gas sector. Pursuing opportunities in this sector has led to the identification of an auspicious application at the fuel blending stage in refineries. At this stage, multiple streams are mixed to achieve the mandated fuel properties based on international and domestic standards. One of the mandated properties is Octane Number (ON). Efforts have been made to optimize the blending operation, nonetheless, machine learning based applications are limited. The intention behind exploring and enhancing this field is to avoid undesirable scenarios, such as off-specification products and quality giveaway. This research presents an innovative approach to predict the optimum fuel composition. This is accomplished through utilizing an integrated system composed of a genetic algorithm (GA) and an artificial neural network (ANN). The integrated system was validated using 201 hydrocarbon, hydrocarbon-ethanol blends and 30 FACEs gasoline-ethanol blends. The system resulted in R^2 of 0.99 for both RON and MON. For the blend compositions, 10 pure components and 6 Fuels for Advanced Combustions Engines (FACE) were evaluated independently. The R^2 value for these data ranged between 0.62 and 0.99. These results signify the success of the integrated system and its major potential impact in mitigating undesirable blending scenarios.

Soot prediction of oxygenated fuels

Mohammed Ameen Mohammed Qasem *a*, Vincent C.O. van Oudenhoven *b*, Eid Al Mutairi *a* and Abdul Gani Abdul Jameel *a*

a Department of Chemical Engineering, King Fahd University of Petroleum & Minerals, Dhahran 31261, Saudi Arabia
b Faculty of Electrical Engineering, Mathematics & Computer Science, Delft University of Technology, 2628 CD Delft, The Netherlands

ABSTRACT

An artificial intelligence (AI) based model using artificial neural networks (ANN) (AI- ANN) was improved for predicting the soot of oxygenated fuels like alcohols and ethers. The hydrocarbon classes such as paraffins, olefins, naphthenes, aromatics, and their mixtures were considered, too. Experimental TSI data of fuels (pure compounds, fuel surrogate mixtures and gasolines) was used as a dataset for developing the AI- ANN model. In this model, ten features (paraffinic CH3, paraffinic CH2, paraffinic CH2, olefinic –CH=CH2, Naphthenic -CH-CH2, Aromatic C-CH, Alcoholic OH, Ether O, molecular weight (MW) and branching index (BI)) have been used as inputs. The composition and structure of the fuel were expressed by the eight mentioned functional groups and the two structural parameters (MW and BI) respectively. The AI- ANN model was trained and finally tested on randomly split sets of the data. The regression coefficient (R2) of 0.97 was observed for test set. The absolute error of prediction of 2.46 was obtained. This is a promising number since it is closed to the uncertainty observed in the experimental measurements. As a result, the fuel's TSI is affected by the fuel functional groups, so it can be used for decision about modeling. Consequently, prediction of TSI of pure compounds, fuel surrogate mixtures and petroleum fuels containing alcohols and ethers can be done by this model (AI- ANN).

Hydrogen sulfide and carbon dioxide removal from natural gas by a robust porous organic polymer

Othman Charles S. Al Hamouz

a Chemistry Department, King Fahd University of Petroleum & Minerals, Dhahran 31261, Saudi Arabia
b Interdisciplinary Research Center for Hydrogen and Energy Storage, King Fahd University of Petroleum and Minerals, Dhahran 31261, Saudi Arabia

ABSTRACT

Natural gas sweetening currently requires multi-step, complex separation processes to remove the acid gas contaminants, carbon dioxide and hydrogen sulfide. In addition to being widely recognized as energy inefficient and cost-intensive, the effectiveness of this conventional process also suffers considerably due to limitations of the sorbent materials it employs. Herein, we report a new porous organic polymer, termed KFUPM-5, that is demonstrated to be effective in the concurrent separation of both hydrogen sulfide and carbon dioxide from a mixed gas stream at ambient conditions. In order to understand the ability of KFUPM-5 to selectively capture these gas molecules, we performed both pure-component thermodynamic and mixed gas kinetic adsorption studies and correlated these results with theoretical molecular simulations. Our results show that the underlying polar backbone of KFUPM-5 provides favorable adsorption sites for the selective capture of these gas molecules. The outcome of this work lends credence to the prospect that, for the first time, porous organic polymers can serve as sorbents for industrial natural gas sweetening processes.

Synthesis of mesoporous Y-Zeolite using Surfactant Templating for Cracking Catalyst Applications

*Adeel Ahmad 1 and Shakeel Ahmed *2*

Chemical Engineering Department, King Fahd University of Petroleum and Minerals,
Center for Refining and Advanced Chemicals, King Fahd University of Petroleum and Minerals,
Dhahran 31261, Saudi Arabia.

ABSTRACT

A wide range of processes and reactions in the industry have utilized zeolites as well as relevant crystalline molecular sieves due to their unique microporous structure, shape selectivity, ion exchange capability, and strong acidity. The small size of cavities and narrow channels inside Y-zeolite impedes their practical applications, and therefore, several studies have been conducted to accommodate the transportation of heavy molecules in the micropore. This study focuses on the generation of mesoporosity inside Y-zeolite, by post-treatment of commercial Y-Zeolite with acid, base, and organic template cetyltrimethylammonium bromide (CTAB). Modified Y-zeolites were used for the preparation of FCC catalysts. Performance evaluation of the catalysts was calculated at 550 °C to see the yield of gasoline, LCO, HCO, and coke. Vacuum gas oil (VGO) was used as feed for testing of FCC catalyst in fixed bed micro activity test (MAT) unit. Modified Y-zeolites were characterized by using XRD, N2 adsorption isotherm, ICP, NH3-TPD and SEM analysis to measure their physical and chemical properties. The relative increase in gasoline selectivity of modified Y-zeolite based FCC catalysts was noted as compared to the base catalyst.